Pat thinks painting is the best thing on earth. She does paintings of her friends at school. She also makes paintings of trees, lakes, and animals.

Sometimes Mom says, "Pat, I think you paint too much. Maybe you should go outside and ride your bike."

So Pat goes outside. But instead of going for a bike ride, she ends up painting!

She paints pretty trees that are greenish-yellow. She paints the sun as a large golden ball. This is a patch of light in the dark blue sky. Several people stop to admire her painting.

When Pat is older, she attends art school. She goes to class from early morning until late afternoon. Her teachers think her work is superb.

Then Pat enters a government contest for young artists. The first prize is a trip to Rome to study in a great art school. Pat is thrilled when she wins first prize!

One year later, Pat comes back to America. She's an even better artist than before. Finally, she sells her first painting! Soon more and more people learn about her work.

Today, many people have heard of Pat. Her paintings can be seen throughout America. They hang in homes, banks, and schools. Pat's hard work has paid off.

Pat usually visits her mom and dad every week. Sometimes she will bring them a new painting. Their house is filled with Pat's work. "Pat was always a real artist," they both smile and say.